Engineering for Inclusion

AA008745

Held at the 2020 AIChE Annual Meeting

Online
16 – 20 November 2020

ISBN: 978-1-7138-2284-4

Printed from e-media with permission by:

Curran Associates, Inc.
57 Morehouse Lane
Red Hook, NY 12571

Some format issues inherent in the e-media version may also appear in this print version.

Copyright© (2020) by AIChE
All rights reserved.

Printed with permission by Curran Associates, Inc. (2021)

For permission requests, please contact AIChE
at the address below.

AIChE
120 Wall Street, FL 23
New York, NY 10005-4020

Phone: (800) 242-4363
Fax: (203) 775-5177

www.aiche.org

Additional copies of this publication are available from:

Curran Associates, Inc.
57 Morehouse Lane
Red Hook, NY 12571 USA
Phone: 845-758-0400
Fax: 845-758-2633
Email: curran@proceedings.com
Web: www.proceedings.com

TABLE OF CONTENTS

596682 Qualitatively Exploring Students' Experiences and Beliefs of Success

Abstract:

Introduction: If the purpose of education is to prepare students to be successful in school and beyond, then we must understand what success is for students. This study aims to fill this gap by understanding senior chemical engineering students' experiences and beliefs of academic success.

Population: Senior chemical engineering students from a large Midwest university were targeted for this study. It is presumed that students on the verge of graduating (seniors), should have experiences academic success at some point in obtaining their degree. Determining how students have experiences academic success is the goal of this study.

Research Question

What are chemical engineering students' experiences and beliefs of success?

Research Design

In order to thickly understand senior chemical engineering students' experiences and beliefs of academic success, a qualitative research design was chosen. An interview protocol was developed following a semi-structured format. The data collected from the interviews will be transcribed and thematically analyzed.

Implications:

If the field of engineering education is to become more inclusive beyond the majority white male student, the experiences and beliefs of minorities must be better understood. Recognizing students' values begins the dialogue of establishing an inclusive culture based on shared values. Educators understanding the range of values articulated by students, may establish their classroom culture with shared values.

Copyright © American Institute of Chemical Engineers. All rights reserved.

Uncovering the Resiliency Strategies and Out-Group Bias for Transgender and Gender Nonconforming Undergraduate Engineering and Computer Science Students

Andrea Haverkamp, Oregon State University, Corvallis, OR

Introduction

The study of gender in engineering continues to be highly relevant due to the persistence of the field's domination by men and masculinity. Mainstream discourse on gender in STEM, however, has been seemingly kept in a "black box" for decades according to writer and researcher Allison Phipps [1]. She states that the reliance on a simplistic gender binary unaccompanied by racial, cultural, or sexual identity nuances undermines engineering's own political aims of gender equity. In other words, all of us are hindered when our conversations regarding gender are simplified to just "men vs. women." The reduction of gender into these two categories has historically invisibilized engineering students whose gender does not exist in those two categories, such as nonbinary or genderqueer students. Additionally, the existing body of gender research in engineering education has not considered gender conformity, gender history, or transgender status as part of their analysis.

We are conducting a multi-phase research project that aims to center the voices and experiences of transgender and gender nonconforming (TGNC) undergraduate, engineering students. The work reported here entails the first-phase of the study, a national outreach questionnaire. The instrument was designed using the Qualtrics platform, and the link distributed through email to undergraduate engineering programs and professional LGBTQ+ engineering organizations nationwide. The objective of the questionnaire was to identify, recruit, and give voice to as many TGNC undergraduate engineering students as possible. The questionnaire consisted of 16 Likert-scale items and 7 open-text prompts, focusing on students' perceptions of their own skills, community support, and personal resiliency strategies. In addition, the questionnaire requested suggestions for ways peers and faculty could improve the learning environment for TGNC students. Demographics on gender, disability, and race/ethnicity were also recorded. These results provide critical insight into the gendered social landscape of engineering for TGNC students.

Responses (n = 353) were separated into two categories: i) 301 respondents were identified as TGNC undergraduate students in engineering education, and ii) 52 respondents were identified as untruthful and malicious repliers. Likert-scale questions were analyzed through descriptive statistics in IBM SPSS for the 301 TGNC responders. The open-word responses were thematically coded in ATLAS.ti for both TGNC and untruthful participants.

Note: For this poster presentation, the following text was provided on the poster in the noted sections. Some of the background text is adapted from prior presentations on this material. Some poster material images were not compatible with this format. Please email the author for copies.

Background

Research documenting or analyzing the experiences of transgender and gender nonconforming

(TGNC) students in engineering is sparse, yet emerging. Insight can be gained from existing LGBTQ+ studies. Lesbian, gay, bisexual, transgender, and queer (LGBTQ+) STEM students in engineering particularly, perceive a lack of respect from their peers while navigating a chilly, discriminatory climate [2-6]. Student experiences of discrimination on the basis of gender and race lead to lower trust in their department and greater disidentification with their programs [7]. A sense of belonging – which can be described as comradery, positive relationships, and identification with peers – is a critical component to persistence in engineering education [8-10].

Studies which seek to describe a universal LGBTQ+ experience may overlook experiences surrounding gender which are specific to the (TGNC) population. Lesbian, gay, and bisexual identities primarily revolve around sexuality and not necessarily gender expression, presentation, and identity. This may inaccurately conflate TGNC gender identities and LGBQ sexual identities as synonymous in experience, despite distinct differences in experience on college campuses [11]. For example, one form of harassment particularly directed towards TGNC individuals is known as "gender bashing" which targets individuals whose gender expressions and identities stray from dominant norms [11-14]. Additionally, TGNC individuals may experience discrimination and exclusion not only from cisgender straight individuals, but from cisgender individuals within the LGBQ+ community as well [11, 12]. These factors result in a need for research focused specifically on the TGNC community in engineering – their experiences, support systems, and pathways to success.

According to the 2015 U.S. National Trans Survey, 24% of people perceived as transgender on college campuses were verbally, physically, or sexually assaulted, and 16% left higher education due to the level of harassment they experienced [15]. These statistics for the overall population are further divided across racial lines, with TGNC individuals of color experiencing greater marginalization. TGNC students who perceive a negative or discriminatory campus climate report decreased measures of student success as measured by performance and retention [16]. The years of higher education are of particular note in TGNC research as nearly half of the US trans population comes out or begins transitioning between the ages of 18 and 24 [15]. This itself is a process with additional social and material support needs which can overshadow the demands of the classroom.

Gender is not a letter on a birth certificate or drivers' license, nor is it simply an internal "identity." Gender is a shifting ever-changing tapestry over all of our lives. It is co-created and replicated daily through countless socio-cultural interactions, consisting of what Judith Butler calls "performance" and the social feedback loop of recognition and validation by others [17,18]. Gender is a shifting phenomenon across era and region, intersecting with race, ethnicity, religion, age, and other identities. This is important as no single chromosomal, hormonal, or psychological factor has been found to be a direct determinant in one's gender identity or expression. Psychological research finds that humans have conceptualizations and expressions of gender which are fluid and unmappable to fixed biological binary, even for cisgender (non-transgender) subjects [19,20]. Instead, the "human brain mosaic" represents fluidity and multiplicity across all humans. Investigating gender in engineering should reflect this nuanced complexity. Studying gender becomes almost academically dishonest when it is reduced to a binary variable which overlooks the identity of nonbinary genders and complexities for trans individuals.

Resiliency framing inspired our study design as opposed to a damaged centered deficit model. Deficit framings of underrepresented groups typically explore *negative* experiences or supposed shortcomings of subordinated groups. A resiliency framework instead investigates strategies of support, success, and daily acts of navigating campus life. The premise is that when you know what helps students succeed and navigate towards success you can institutionally and culturally center and strengthen the programs or structures identified. Asking TGNC students for their ideas on increasing inclusivity based upon their own daily experiences, rather than focusing on harm, will produce readily accessible results. By studying the way that those with sharp experiences of invisibility or marginalization succeed in engineering we can uncover fundamental insight into how gender shapes culture and climate for all of us.

Methodologies

The study uses community-based methodologies as informed from both engineering education and trans studies. Our approach synthesizes these two fields together for a unique approach to the study of TGNC students. In particular, feminist ways of knowing and community centered research are the basis of our study design.

We recognize the challenges to positivist notions of scientific objectivity when investigating identity and the inherently affective experiences of belonging, safety, and success. One aspect of implementing this awareness into the research process is understanding ourselves through subjective lenses. Recognizing the positionality of our identity, experience, and power in the academy can shed light on implicit biases. This helps form "reflexivity" in the research process, an iterative cycle of self-analysis and change in approach as results begin to be recorded [21]. Feminist standpoint theory has been applied in prior engineering education research to demonstrate that the experiences to be studied are best understood by the participants themselves [22]. Standpoint theory places the research participant as the subject matter expert on their own lives. For reflexive research that recognizes the subjective researchers' positionality, we then need to bring the research participant to the research table as part of the analysis and iterative changes to the study design.

We position TGNC students as the foremost experts on their experiences, and collaborate with TGNC individuals in every phase of the research. As such, it is necessary to involve TGNC students at every phase of the research. Involving the research subjects as part of the research team is integral to participatory and community collaborative research methodologies, such as those used by Z. Nicolazzo in their research with undergraduate trans-identified students [14]. These methods involve the subject community as much as possible at every phase of the research, from data collection, collaborative analysis, authorship, and informing mutually beneficial end products. Research Justice as an academic movement inspired us to ensure that our participants obtained fair and just compensation for their time and effort at every part of their interaction with the research [23].

Data Collection

In order to address these issues, we are conducting a multi-phase research project that centers the voices and experiences of transgender and gender nonconforming (TGNC) undergraduate engineering students. The work reported here entails the first-phase of the study, a national outreach questionnaire. The instrument was designed using the Qualtrics platform, and the link distributed through email to undergraduate engineering programs and professional LGBTQ+ engineering organizations nationwide. The objective of the questionnaire was to identify, recruit, and give voice to as many TGNC undergraduate engineering students as possible. The questionnaire consisted of 16 Likert-scale items and 7 open-text prompts, focusing on students' perceptions of their own skills, community support, and personal resiliency strategies. In addition, the questionnaire requested suggestions for ways peers and faculty could improve the learning environment for TGNC students. Demographics on gender, disability, and race/ethnicity were also recorded. These results provide critical insight into the gendered social landscape of engineering for TGNC students.

Responses (n = 353) were separated into two categories: i) 301 respondents were identified as TGNC undergraduate students in engineering education, and ii) 52 respondents were identified as untruthful and malicious repliers. Likert-scale questions were analyzed through descriptive statistics in IBM SPSS for the 301 TGNC responders. The open-word responses were thematically coded in ATLAS.ti for both TGNC and untruthful participants.

Results

Resiliency

For this poster we focus on one prompt related to resiliency: "How can engineering students better support one another within their major and their program?" The responses are rich with ideas on how TGNC students know how their experiences could be improved, and persistence supported daily. The suggestions are a critical early glimpse into the ways researchers and educators can improve our programs. We present quotes from students in this poster to support the major themes of cultivating a more welcoming and friendly engineering cultural climate, providing social justice and gender related education on identity, and paying attention to classroom inclusivity.

The most prevalent themes were:

- Cultivating friendliness, respect, open-mindedness, replacing competition with collaboration, and developing healthy interpersonal communication.
- Creating social awareness through education. Many further noted that engineers were not equipped with education or a culture that fostered awareness of trans identities or LGBTQ issues, corresponding to response themes for educators to create social awareness through education on social justice.
- Building or developing safer and more inclusive physical spaces, and to have their peers educated on pronouns and trans identities.
- Bystander intervention when hate or bias occurs: a number of responses described negative experiences or a sense that engineering and computer science undergraduate student culture needs a dramatic change. Respondents frequently noted that they had to argue the validity of trans peoples' existence or heard "jokes" which diminished LGBTQ+ people. The students' other identities such as disability and race were mentioned alongside gender as they wrote about experiences of ableism, classism, and racism. A few went further to specifically describe a culture of toxic masculinity which hurt all gender minorities.

Peer Bias

The following sections outline major patterns found in the 51 bias responses. Overall, we believe the wide range of bias presented as backlash to our research questionnaire demonstrates the *intersectionality of oppression* - i.e. bias against trans individuals is not isolated, but is instead related to a wide matrix of dominance and oppression.

- Politics of Diversity - the biased responders frequently lamented the perceived pervasiveness of diversity quotas, affirmative action, and social justice "privilege point systems." Programs which seek to promote equity and inclusion may be perceived by these respondents as antithetical to ideologies of meritocracy within engineering and computer science.

- Gender Binary Discourse - Respondents communicated their belief in an immutable biological gender binary both overtly and tangentially through their commentary on contemporary theorizations of gender. The most overt was one respondent putting the exact wording "There are only two genders" into every open-text comment field, echoed by similar simple wording provided by others.

- Transphobia - As could be reasonably predicted, a lot of the comments targeted TGNC individuals and their community. Trans and queer people were referenced with hostility, anger, resentment, denial of their identity's validity, and tropes such as pedophilia and sexual predatory behavior.

Demographic gender identities given by biased responders range from those identical to valid TGNC respondents (e.g. trans male, nonbinary queer, female, agender) to those clearly indicated to be an intentional distortion of TGNC identity (e.g. Non-cookie-cutter cis-furry dragonkin, homophobic bigot, non-binary intersectional African Mohammedian Feminist, pedophile).

- Racism - Racial and ethnic identity demographics full of hateful language show that race, racism, and race theory are deeply embedded throughout the data. Especially anti-Blackness. It is not enough to simply list "Black", but Blackness was compounded with additional terms to form identities such as "Black Hispanic Jew", "Blasian", "Hispanic latina, native american black", "Black, Hispanic American, Hawaii Indian, "African Jewish", and "Black and native American." It is the mixing of races and ethnicities that is most pronounced in the data signaling one of white supremacy's primary reactionary fears, which is a diluting of whiteness. Anti-Black online netspeak which serves as a reference only to those in the hate community was found here as well. A fixation on mocking Indigeneity was present as well.

- Anti-semitism - Comments throughout the questionnaire signaled white supremacy not only through racism, but also anti-Semitism.The Jewish graduate researcher whose name was listed on emails and consent forms was also referenced directly as well.

Next Steps

The outreach questionnaire provided the research team with an initial view into the lives, identities, and success strategies of TGNC engineering and computer science students. These results have been used to inform subsequent research activities including phone conversations and in-person ethnographic site visits, both of which provide deeper, complimentary, and context-specific narrative data.

References

1. Phipps, A. (2007). Re-inscribing gender binaries: Deconstructing the dominant discourse around women's equality in science, engineering, and technology, *The Sociological Review, 55*(4), 768-787.

2. E. Cech, T. Waidzunas, STEM Inclusion Study Organization Report: APS. Ann Arbor, MI: University of Michigan. Report. 2018.

3. E. Cech, T. Waidzunas, "Navigating the heteronormativity of engineering: The experiences of lesbian, gay, and bisexual students," Engineering Studies, vol. 3, no. 1, pp 1-24. 2011.
4. E. Cech, T. Waidzunas, "Navigating the heteronormativity of engineering: The experiences of lesbian, gay, and bisexual students," Engineering Studies, vol. 3, no. 1, pp 1-24. 2011.
5. J. Yoder, A. Mattheis, "Queer in STEM: Workplace Experiences Reported in a National Survey of LGBTQA Individuals in Science, Technology, Engineering, and Mathematics Careers. Journal of Homosexuality, vol. 63, no. 1, 2016.

6. K. Trenshaw, A. Hetrick, R. Oswald, S. Vostral, M. Loui, Lesbian, Gay Bisexual, and Transgender Students in Engineering: Climate and Perceptions: Proceedings – Frontiers in Education Conference, October 2013.

7. C. Baker, "Social support and success in higher education: The influence of on-campus support on African American and Latino college students," The Urban Review, vol. 45, no. 5, pp 632–650. 2013.

8. K. Meyers, M. Ohland, A. Pawley, S. Silliman, K. Smith. "Factors relating to engineering identity," Journal of Engineering Education, vol. 14, no. 1, pp 119-131. 2012.

9. D. Means, C. Beatty, R. Blockett, M. Bumbry, R. Canida, T. Cawthon, "Resilient Scholars: Reflections from Black Gay Men on the Doctoral Journey," Journal of Student Affairs Research and Practice, vol. 54, no. 1, pp 109-120. 2017.

10. E. Tate, M. Linn, "How Does Identity Shape the Experiences of Women of Color Engineering Students?" Journal of Science Education and Technology, vol. 14, no. 5-6, pp 483-493. 2005.

11. J. Dugan, M. Kusel, D. Simounet. "Transgender College Students: An Exploratory Study of Perceptions, Engagement, and Educational Outcomes," Journal of College Student Development, vol. 53, no. 5, pp 719-736. 2012

12. D. Hill, "Genderism, transphobia, and gender bashing: A framework for interpreting anti-transgender violence," in Understanding and dealing with violence: A multicultural approach, B. Wallace and R. Carter, Eds. Thousand Oaks, CA: Sage, 2002, pp. 113-136.

13. J. Weiss, "Reflective Paper: GL Verses BT: The Archaeology of Biphobia and Transphobia Within the U.S. Gay and Lesbian Community," Journal of Bisexuality, vol. 11, no. 4, pp. 498-502, 2011.

14. J. Weiss, "Reflective Paper: GL Verses BT: The Archaeology of Biphobia and Transphobia Within the U.S. Gay and Lesbian Community," Journal of Bisexuality, vol. 11, no. 4, pp. 498-502, 2011.

15. S. James, J. Herman, S. Rankin, M. Keisling, L. Mottet, and M. Anafi, "The Report of the 2015 U.S. Transgender Survey," National Center for Transgender Equality, Washington D.C. Report. 2016.

16. J. Pryor, "Out in the Classroom: Transgender Student Experiences at a Large Public University," Journal of College Student Development, vol. 56, no. 5, pp 440-455. 2015.

17. Butler, J. (1990). Gender Trouble: Feminism and the Subversion of Identity. Abingdon, UK: Routledge.

18. Langer, S.J. (2016). Trans Bodies and the Failure of Mirrors. Studies in Gender and Sexuality, vol. 17(4), pp. 306-316

19. Joel, D., Tarrasch, R., Berman, Z., Mukamel, M., Ziv, M. (2014). Queering Gender: studying gender on gender-normative individuals, Psychology & Sexuality, vol. 5, no. 4, pp. 291-321.

20. Joel, D., Berman, Z., Tavor, I., et. al. (2015). Sex beyond the genitalia: The human brain mosaic. PNAS, 112(50), pp. 15468-15473.

21. Borrego, M., Douglas, E., Amelink, C. (2009). Quantitative, Qualitative, and Mixed Research Methods in Engineering Education. Journal of Engineering Education, 98(1), 53-66.

22. A. Pawley, A. (2013). Learning from small numbers" of underrepresented students' stories: Discussing a method to learn about institutional structure through narrative," 120th American Society for Engineering Education Annual Conference & Exposition, Paper ID# 6639.

23. Jolivette, A. (2015). Research Justice: Methodologies for Social Change. Policy Press, University of Bristol.

608603 Diversifying the U.S. Professoriate: How to Expand the Academic Pipeline?

Abstract:

Diversifying the U.S. professoriate today is essential for the intellectual health of higher education. Diversity of a university's students and faculty affects its strength and efficiency. Research shows that a diverse student body and faculty contributes to their university's creative and innovative ecosystem[1]. Yet diversity brings its own set of challenges. Underrepresented minority students (URMs) continue to struggle with identity, isolation, and representation, and too often, leave their academic environment[2].

According to diversity research[3-4], role-models and mentors who share cultural/racial/socio-economic/ethnic identities contribute to their students' desire to pursue similar career paths and motivate them to remain resilient. As a minority student in U.S. universities and as a beneficiary of excellent mentors in STEM fields, I wouldn't have been able to overcome my challenges and pursue graduate school at Stanford without the support of my mentors. My unique experiences as an international exchange student from Pakistan at the University of Mississippi several years ago and as the only Muslim woman in my current PhD cohort have enabled me to understand some of the socio-economic, religious, and cultural struggles of women and URMs in STEM fields. My experiences also exemplify the power of diversity in U.S. academia and the need to create an academic culture where all students feel comfortable.

During my presentation, I will talk about some of these experiences and share my insights with fellow students to help them navigate their own academic journeys. I will also discuss specific examples of being mentored that will help other mentors/advisors show sensitivity to the barriers faced by their URMs. As an aspiring faculty member, I plan to implement these best practices in my own research laboratory one day and continue advocating for URMs in the scientific community.

References:

1. Women in Science & Engineering Leadership Institute, "Benefits and Challenges of Diversity in Academic Settings,": https://wiseli.wisc.edu/wp-content/uploads/sites/662/2018/11/Benefits_Challenges.pdf .
2. Brunsma, David L., David G. Embrick, and Jean H. Shin. "Graduate students of color: Race, racism, and mentoring in the white waters of academia."*Sociology of Race and Ethnicity* 1 (2017): 1-13.
3. Ebony McGee, "Mentoring Underrepresented Students in STEMM: A Survey and Discussion," Committee on Effective Mentoring in STEM: https://www.nap.edu/resource/25568/McGee%20-%20STEMM%20Mentoring%20Identity.pdf .
4. Aikens, Melissa L., et al. "Race and gender differences in undergraduate research mentoring structures and research outcomes."*CBE—Life Sciences Education* 2 (2017): ar34.

Copyright © American Institute of Chemical Engineers. All rights reserved.

595502 Using a Diverse, Inclusive and Supportive Academy Integrating Research and Professional Skill Development to Enhance Graduate Education in Chemical Engineering

Abstract:

Graduate education and training often take a monodisciplinary approach that is not informed by best educational practices, ignores the needs and preferences of trainees in general and underrepresented minority (URM) trainees in particular, and overlooks the increasingly interdisciplinary and international nature of research. In this contribution, we describe a diverse, inclusive and supportive academy integrating research and professional skill development to enhance graduate education in chemical engineering. This academy – which leverages the most effective tools for the training of chemical engineering professionals (including developmental network-based mentoring and career-transferable skill development) – will generate an innovative model for graduate student training by identifying and avoiding issues often associated with these tools.

For instance, mentoring has been proven to improve retention and performance, leadership, and student involvement in graduate school, especially among URM students. However, mentoring is typically challenged by the lack of adequate training not only for mentors but also for mentees, an issue only recently emphasized by mentoring experts. Similarly, although the skills that graduate students need are well known, students are commonly left to drive their own development, with "sink or swim" constituting the default advising style. Several theories exist regarding the mechanism through which mentoring improves retention and academic success, including promoting students' involvement in learning, facilitating their integration in the educational environment, helping them cope with this environment and providing the developmental support they need to reach their potential. Since using a single theory is inadequate due to various possible outcomes, all theories aligned with the outcomes pursued must be acknowledged. Indeed, each theory emphasizes distinct mentoring relationships given that different types of mentors are best at providing specific kinds of support. Thus, multiple relationships are required to ensure the attainment of the desired outcomes, which highlights the need for – and the value of – a developmental network approach to provide trainees with all the support they need. This approach is particularly effective at improving outcomes for URM students.

To retain URM students, universities must provide a supportive and nurturing environment, mentoring also helping URM students succeed in an environment that may be alienating or hostile. Mentoring has even been defined as "a deliberate effort to provide support to minority students [...] through frequent contact and interactions with mentors" [N.H. Cohen, 1995]. Notably, mentoring has been successfully used to attract and improve the retention of URM students, who found mentors of great help to their academic and professional development. This is particularly true for URM students in STEM, for whom a close student-mentor relationship and role modeling lead to positive outcomes. Among the literature on the mentoring of STEM students, reports of increases in the number of URM students who earned PhDs are noteworthy. Nevertheless, the issue of underrepresentation in STEM at the graduate level and in the workforce remains ingrained.

Attrition among URMs has been attributed to academic and cultural isolation, lack of peer support, low motivation due to low expectations, and discrimination. All these issues can be addressed by mentoring, which can increase retention and the number of students who earn STEM baccalaureate and graduate degrees. Recruiting active research faculty and involving students in research represent two important elements to improve academic integration, knowledge and skill development, as well as to offer support and motivation. This renders a research and developmental network-based mentoring approach – and its alignment with the aforementioned mentoring elements and objectives – particularly compelling. Last, while some mentoring programs pair students with a mentor of the same sex or ethnicity, many reports emphasize that cross-race and/or gender mentoring is effective and offers distinct advantages to mentees, including improving cross-race and cross-gender communication skills. Moreover, reports indicate that academic interests are an equally sensible matchmaking parameter and that matching by gender or race does not improve outcomes, which is propitious since same-race or same-gender matches are challenging at the graduate level where, by definition, faculty from underrepresented groups is scarce.

It should also be recognized that a one-size-fits-all approach can potentially perpetuate discrimination for URMs unless intersectionality – a term originally coined to discuss instances "where women of color experience *both* race and gender discrimination simultaneously" [V.H. Hunt *et al.*, 2012] – is addressed. Thus, students must be treated as diverse individuals with multiple intersecting identities and target the recruitment, retention and mentoring of women of color by recruiting at minority serving institutions, considering the impact of application fees or minimum GPA and GRE scores, and identifying sources of financial aid. Critical multicultural education and intersectionality must be used as a combined framework to help all participants understand cultures other than their own.

Against this backdrop, the training graduate students receive – in a way that is fully integrated with the research they perform – within the diverse, inclusive and supportive academy described herein includes: 1) tools such as individual development plans and developmental network maps; 2) a multi-departmental and interdisciplinary course on research-related content; 3) a seminar course on transferrable skills (ethics, research, communication, teaching, mentoring, entrepreneurship, teamwork, management, leadership, outreach, etc.); 4) a certificate to be awarded once students complete the two courses above and garner additional credits from an interdisciplinary curriculum of research-related courses; 5) summer internships at other departments and at external institutions (other universities, industry, national laboratories) nationwide or abroad; 6) an annual research-related symposium including all elements of a scientific conference; 7) internal collaborative research grants for participants to fund and pursue their own ideas; 8) fields trips to facilities related to the research; and 9) coaching on job hunting as well as résumé, motivation letter and interview preparation.

Since a workforce equipped to meet society's challenges must be both well trained and diverse, multiple initiatives ensure that the academy broadens participation in chemical engineering. Recruitment-wise, close collaboration with a number of entities provide the academy with a broad recruitment pool of talented and diverse students. Moreover, collaboration with these entities provide trainees with ample opportunities to acquire, practice and refine their professional skills, as trainees present their results and recruit in conferences, meetings and outreach events organized by these entities, become members and/or join their leadership, and expand their professional and mentoring network in the process. In addition, URM trainees must be surveyed periodically to probe their feelings of well-being, preparation, acceptance, belonging and distress, as well as their perception of how well structured their departments and programs are. According to recent literature, these factors determine whether or not they perform (i.e., publish) at rates comparable to their male majority peers.

Saliently, the evaluation of the educational model employed is designed to afford a comprehensive understanding not only of the academy components that were more utilized and impactful, but also reveal the individual mentoring and skill-building facets of the program driving its successful implementation. The evaluation plan includes outcomes, performance measures, an evaluation timetable, benchmarks and a description of how formative evaluation will improve practice, the evaluation process also extending to research activities.

Copyright © American Institute of Chemical Engineers. All rights reserved.

The National Academies of SCIENCES ENGINEERING MEDICINE

Ashley Bear, Ph.D.
Senior Program Officer
Committee on Women in Science, Engineering, and Medicine

Presentation for American Institute of Chemical Engineering Annual Meeting 2020

"Status of Women and Men in Science and Technology Professions"

The National Academies of SCIENCES ENGINEERING MEDICINE

The National Academies of SCIENCES ENGINEERING MEDICINE

Available for free at nap.edu

The National Academies of SCIENCES ENGINEERING MEDICINE

abear@nas.edu

The National Academies of SCIENCES ENGINEERING MEDICINE

Three Key Takeaways

1. Women—especially women of color—are underrepresented in science and engineering, but their representation varies by discipline

2. Women's underrepresentation is driven in large part by bias, discrimination, and harassment

3. There are strategies and practices that institutions have used to improve this situation

The National Academies of SCIENCES ENGINEERING MEDICINE

Representation Varies by Discipline and Career Stage

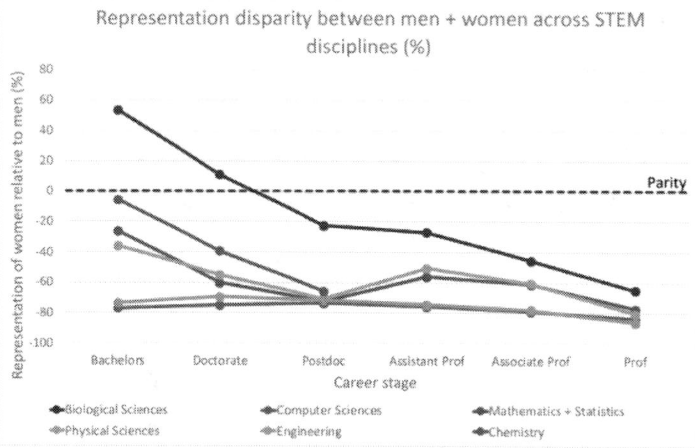

Women of Color Remain More Underrepresented than White Women

Science and engineering degrees earned by underrepresented minority women and men: 1995–2014

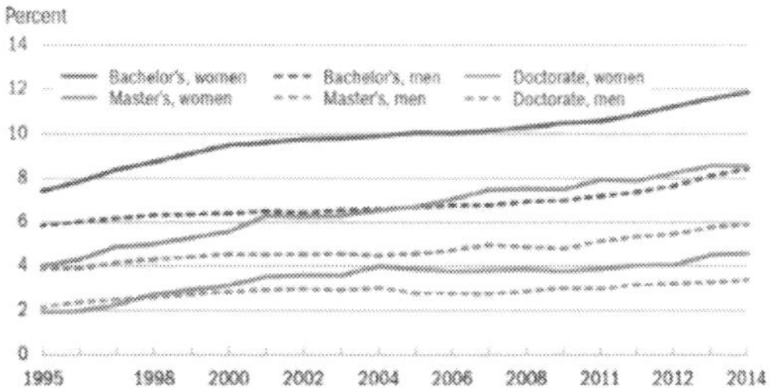

NOTE: Data not available for 1999.
Women, Minorities, and Persons with Disabilities in Science and Engineering: 2017

The National Academies of SCIENCES ENGINEERING MEDICINE

Why Are Women Underrepresented?

Bias, discrimination, and harassment are major drivers of the underrepresentation of women in science, engineering, and medicine

Often experienced more overtly and intensely by women of multiple marginalized identities

Women of color also experience isolation; exclusion from social network supports; a sense of not belonging; cumulative disadvantages in pay and promotion that accrue over time; and racial harassment

The National Academies of SCIENCES ENGINEERING MEDICINE

Why Are Women Underrepresented?

Obtaining a Position—bias in recruitment; obstacles to accommodating family needs

Internal Opportunities and Rewards—unequal allocation of resources; mentoring and performance evaluation (e.g., teaching); mentoring access

Work Expectations—higher teaching loads; higher expectations of service without compensation

External Opportunities and Rewards—lower frequency of speaking invitations; inequities in access to external funds; less likely to be on editorial boards and in editor positions

The National Academies of SCIENCES ENGINEERING MEDICINE

Biases and Barriers Manifest in Different Ways in Different Disciplines

Biology and Medicine

- "hustle" culture that encourages work-life blurring
- commitment to job
- ideal worker norm

Engineering, Computer Science, and Physics

- "innate brilliance"
- lack of role models and mentors
- gender disparities in pay
- lack of recognition and advancement opportunities

The National Academies of | SCIENCES ENGINEERING MEDICINE

Prevalence of Sexual Harassment

Sexual harassment is common in academic science, engineering, and medicine.

- Best available analysis to date shows that 50 percent of women faculty and staff in academia experience sexual harassment.
- Student surveys at a couple universities shows that between 20-50 percent of students in science, engineering, and medicine experience sexual harassment from faculty or staff.

The National Academies of | SCIENCES ENGINEERING MEDICINE

Three Forms of Sexual Harassment

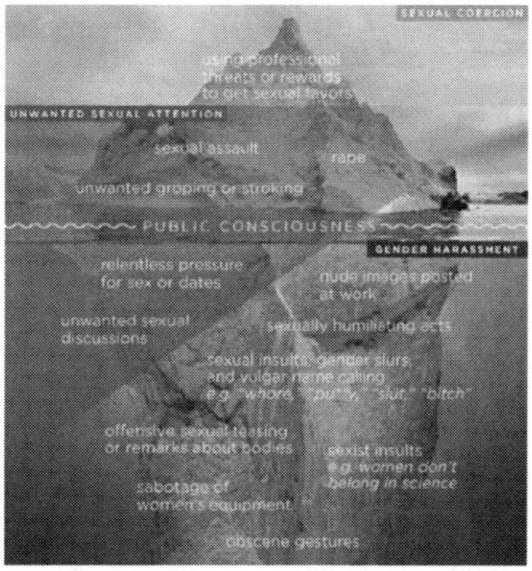

The National Academies of | SCIENCES ENGINEERING MEDICINE

What are Possible Solutions?

Finding: A system-wide change to the culture and climate in higher education is required to prevent and effectively address all three forms of sexual harassment

Recommendation: Create diverse, inclusive, and respectful environments

The National Academies of | SCIENCES ENGINEERING MEDICINE

What are Possible Solutions?

Educational approaches

Active learning, growth mindset, changing the image of science and engineering fields, role models, mentoring

Practices that can support improved recruitment, retention, and advancement of women in these fields

Ways to expand networks of job candidates, guidance on how to write job advertisements and conduct interviews inclusively, strategies to mitigate bias in hiring and promotion decisions, resources that support the work-life needs of professionals and students

The National Academies of | SCIENCES ENGINEERING MEDICINE

Why Don't We See More Progress?

Research points to a common set of conditions that support successful institutional adoption of practices to improve the recruitment, retention, and advancement of women, including:

- **Committed leadership at all levels**
- **Dedicated financial and human resources**
- **Accountability and data collection**
- **Adoption of an intersectional approach**

The National Academies of | SCIENCES ENGINEERING MEDICINE

Sponsors

NSF NIH

BURROUGHS WELLCOME FUND

HENRY LUCE FOUNDATION

NASA NIST NOAA

hhmi Howard Hughes Medical Institute

Alfred P. Sloan FOUNDATION

618541 Metabolic Modeling of *Pseudomonas putida* KT2440 tounderstand and improve the breakdown of plastic waste

Abstract:

Polyethylene terephthalate (PET), the polymer used to manufacture plastic beverage bottles, is the world's highest production-volume synthetic plastic. While synthetic polymers are typically resistant to biodegradation, organisms have been discovered that can decompose PET into soluble but toxic monomers, ethylene glycol (EG) and terephthalic acid (TPA). *Pseudomonas putida* is a soil bacterium known for its receptiveness to genetic engineering and ability to metabolize a wide assortment of organic compounds and pollutants. We seek to understand the pathways by which *P. putida* can metabolize EG and/or TPA and whether its growth on these compounds could be improved by genetic modification. Our analyses are based on whole-genome metabolic models of *P. putida* accessed with the web platform KBase and locally implemented in a Python programming environment, COBRApy. These models are used to generate flux balance analysis (FBA) outputs that allow us to quantitatively predict the organism's growth in different media. FBA results indicate that *P. putida* KT2440, described by the model iJN1463, is capable of biomass formation on EG. We are currently investigating growth on TPA. Our research has uncovered metabolic "bottlenecks" that, if alleviated by genetic engineering, could improve the ability of *P. putida* to metabolize these carbon sources.

Copyright © American Institute of Chemical Engineers. All rights reserved.

618544 Characterization of Rice Residues for Biogas Production by Co-digestion with Goat Manure

Abstract:

Anaerobic digestion is the primary technique used in aiding and limiting odors in waste management systems, reducing the organic strength in waste streams, and is a major player in producing a high energy source biogas. The anaerobic digestion process is a process of degradation of organic materials by the microorganisms that operate in an oxygen-free environment. This work focuses on the anaerobic co-digestion of goat manure (GM) with rice residues (straw and husk) to determine the maximum biogas yield through crop residues. Goat manure was collected from International Goat Research Center, PVAMU. Rice residues were collected from Gulf Pacific Co. (Houston, TX) and passed through Willy mill, subsequently sieved (2 mm) to obtain uniform sized particles. The fresh inoculum was obtained from the Prairie View wastewater treatment plant (operating at mesophilic temperature) and was de-gassed for a week. The well-mixed samples in triplicates were characterized for moisture, total and volatile solids (TS and VS) and the elemental composition (C, H, N, and S). The C/N ratio of the RS and RH was 66.08 and 66.51, respectively. The volatile solid content of the residues was about 80%. The future experiments will focus on determining the best mixing ratio for co-digestion of GM with rice residues and fabrication of two-liter lab-scale bioreactors. The biodegradability and the kinetic parameters will be determined in from the daily biomethane data and biomethane potential (BMP), respectively.

Copyright © American Institute of Chemical Engineers. All rights reserved.

618545 The Pathophysiological Effects of Fluid-Structure Interaction of Species Transported and Transformed from Ambient to Human Respiratory System

Abstract:

The physiological system in humans consist of several organs that are structured to relate chemically and physically with complex fluids which aid to carry different nutrients and gaseous species (oxygen) necessary for proper functioning of the human system. However, the quality of the ambient atmosphere and its physiological end effect plays a tremendous role in determining homeostasis (stability) of the human body systems. For instance, anthropological alteration of the quality of air such as incomplete combustion of hydrocarbon can lead to release of injurious gases such as carbon monoxide and if inhale by human can distort the composition of blood leading to the formation carboxyhemoglobin. This physiological distortion is fatal in its consequence. Furthermore, with the increased interest in correlating ambient fluid quality with human pathophysiological conditions, it has become immensely significant to understand the mechanism of species transported and transformed via the respiratory tract, the major gateway to human complex systems. Thus, the research is aimed at studying the pathophysiological impact of ambient species on human physiological system through detail analysis of the flow process of species and its mechanism of interaction as they are transported from the bulk ambient fluid to the respiratory organs. A detail qualitative study will be conducted to gather data on the composition of the ambient atmospheric fluids at different geographical locations and coupled these data into modelling platform designed to study the mechanism of transportation and transformation of species as they interact with the organs of the respiratory system. The goal of this work is to develop a prognosticative and diagnostic modelling platform that can successfully analyze and predict the pathophysiological effects of ambient fluid on human systems while offering suggestions on range of preventive and therapeutic measures to correct these physiological anomalies.

Copyright © American Institute of Chemical Engineers. All rights reserved.

Olefins Plant Optimization using SFT Technology in cracking furnaces

[1]Karen Romero Harrington, [2]Daniel Salerno, [3]Jorge Sánchez

[1,3] Universidad del Zulia. [2]UNET

[1]karenrosana@gmail.com, [2]danielsalerno@gmail.com, [3]jlsanchez@fing.luz.edu.ve

Abstract— The thermal cracking of hydrocarbons in the presence of steam is the key unit of an olefins plant. Ethylene and propylene are basic materials for obtaining a great important variety of petrochemical products for diverse applications. The feedstock costs fluctuation and high demand of products require the optimal operation of the reactor for hydrocarbon pyrolysis. A novel reactor technology for steam cracking furnaces, called Swirl Flow Tube (SFT) as an alternative to improve furnace operation, has been evaluated computationally using ANSYS Fluent, based on a detailed free-radical mechanism. Cracking furnaces of an existing olefins plant were used to study this technology due to its low production and operational failures because of coke formation leading damages to the material of the coils.

They present $6.2 million per year losses in ethylene revenues and up to $70 million per year in OPEX costs. A computational fluid dynamics model (CFD) was adopted to capture the main flow properties of the swirl flow tube. The economic evaluation was carried out using Aspen Process Economic Analyzer, considering CAPEX and OPEX estimates, comparing "straight" tubes (original design) vs. SFT-M and SFT-H. The simulation results confirm the great potential for the application of SFT technology. Incorporating SFT-M ("the best alternative") would reduce $0,6 per ton of ethylene in utility costs and 17.5% in operational and maintenance costs compared to "straight" tubes. It guarantees operational continuity, increases ethylene production of up to 13.5%, reduces OPEX by $0.6 per ton of ethylene and rises furnace run length of up to 80 days because of reducing coke formation, without affecting mechanical integrity of coils. The SFT-M CAPEX contemplates $577,823.00, which would be retrieved in olefins production in a period no longer than 6 months.

Index Terms—: Pyrolysis furnaces, tubular reactor, reactor modeling, olefins production, Swirl Flow Tube

I. INTRODUCTION

Production of olefins is one of the main processes in petrochemical industries. Thermal cracking of hydrocarbons is the main alternative for obtaining ethylene and propylene.

The current demand for petrochemical products has significantly increased, so it is mandatory to create a precise model to reproduce operational conditions and enhance the energy efficiency of the process to maximize production with low energy and utility requirements. Being the heart of an olefins plant, furnace technology continues to be a dynamic area of research.

Undesirable reactions that influence process operation and profit margins are CO formation and coke deposition. Coke layer inhibits heat transfer from the tube to the process gas, and hence is responsible for the rising tube wall temperatures

These affect furnace-cracking profit in 3 routes: increased energy consumption, loss of furnace availability because of decoking/mechanical failure and the lost in olefin selectivity due to an increment pressure drop. Multiples researching have been done in this area during the last 30 years to find adequate methods to reduce coke formation. These technologies can be classified into three types: the use of additives, metal surface technologies and the use of mechanical devices to improve heat transfer. In the last class, three-dimensional reactor designs are used to improve heat transfer [1], resulting in lower wall temperatures and coking rates. A novel reactor technology for an existing furnace cracking, called Swirl Flow Tube (SFT), has been evaluated experimentally and computationally. Cracking furnaces of an existing olefins plant (industrial furnaces) were used to study this technology due to its low production and operational failures because of coke formation leading damages to the material of the coils. Three geometries were used to evaluate feasible of this technology: straight tubes (original design), mild swirl flow tubes (SFT-M) and high swirl flow tubes (SFT-H).

The economic evaluation was carried out using Aspen Process Economic Analyzer, considering capital cost (CAPEX) estimates and operating cost (OPEX) estimates, comparing "straight" tubes vs. SFT-M and SFT-H.

II. PROCESS DESCRIPTION

The ethane and propane cracking furnaces of the existing olefins plant are composed of convection and radiation sections as depicted in Fig. 1. The feed is introduced into the convection section. This stream is diluted with steam (0.35 kg steam/kg ethane or kg propane). The pyrolysis reactions continue in tubular coils made of Cr/Ni alloys. Then, this mixture is preheated to around 500–800 °C by using hot stack gas. After that, it goes into the radiation section, where ethane/propane molecules break up to produce ethylene, propylene, butadiene, methane and other products. This procedure takes place in radiant tubes (4-2-1 configuration) that are located vertically in a rectangular furnace at 800–900 °C. Forty side burners (160 burners) provide the required heat to maintain the reactor temperature at this range. In this process, temperature role is of paramount importance for output concentrations and so the coil outlet temperature (COT) of these furnaces should be kept constant between 830°C-860°C.

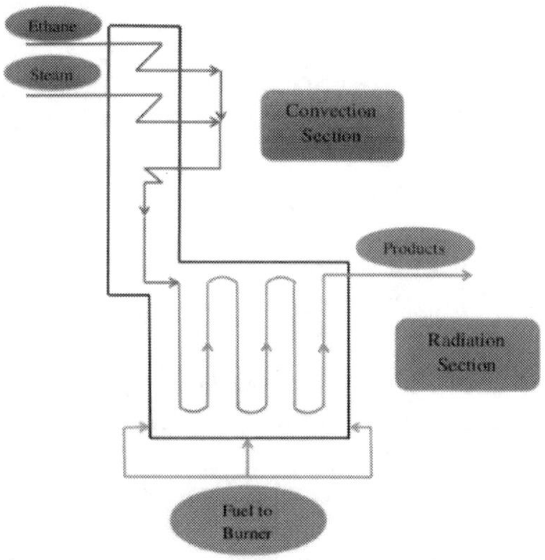

Figure 1. Process diagram furnace in thermal cracking

III. EQUATIONS MODEL

The free radical reaction network suggested by Sundaram and Froment [2] was used to model the pyrolysis of ethane and propane. Continuity equations for the various process gas species area solved simultaneously with the energy, momentum and the coking rate equations required. These equations are as follows:

Mass balance:

$$\frac{dF_j}{dx} = \left(\sum \alpha_{ij}.r_i\right)\left(\frac{\pi d_t^{\,2}}{4}\right) \quad (1)$$

Energy balance:

$$\textstyle\sum_j F_j\, C_{pj} \frac{dT}{dx} = Q(x)\pi d_t + \frac{\pi d_{in}^{\,2}}{4} \sum_i r_{ri}(-\Delta H)i \quad (2)$$

Momentum balance:

$$\frac{dP}{dx} = \frac{\frac{\sum_j \frac{dFj}{dx}}{G_m.A_g} + \frac{1}{M_m}\left(\frac{1dT}{xdx} + F_r\right)}{\frac{1}{PM_m} - \frac{g_C.P}{\rho^2 R.T}} \quad (3)$$

with

$$F_r = 0.092\frac{Re^{-0.2}}{d_{in}} \quad (4)$$

For the straight parts of the reactor coils:

$$F_r = 0.092\frac{Re^{-0.2}}{d_{in}} + \frac{\xi}{\Pi.R_b} \quad (5)$$

The parameter of tube bend (ξ) is expressed as follows:

$$\xi = \left(0.7 + 0.35\frac{\Lambda}{90°}\right).\left(0.051 + 0.19\frac{d_{in}}{R_b}\right) \quad (6)$$

Heat transfer model

$$Nu = \frac{\phi Cp.(T_{out} - T_{in})}{\pi.\lambda.L_h.LMTD} \quad (7)$$

Coke formation model

The rate of coke formation based on Sundaram et al (1981) investigation [3], was calculated from the equation:

$$\Delta t_c = \frac{(x_1.r_{c1}.C_{C4} + x_2.r_{c2}.C_{C3H6}).\Delta T_{time}}{\rho_c.10^6} \quad (8)$$

After each time interval ΔT_{time}, the tube diameter was update by:

$$Dt_{,new} = Dt_{,old} - 2\Delta T_c \quad (9)$$

To find out the best-operating conditions for cracking furnace of an existing olefins plant, the furnace was modeled using Ansys Fluent software. This general-purpose CFD package uses the finite-volume method.

To suitably resolve the viscosity affected near-wall region, a two-layer model was adopted. In this model, the domain is subdivided into a viscosity-affected region and a fully turbulent region by evaluation of a wall-distance-based turbulent Reynolds number, defined as:

$$Re_y = \frac{\rho.y.\sqrt{k}}{\mu} \quad (\mathrm{Re}_y > 200) \quad (10)$$

where y is the distance between the cell center and the nearest wall cell.

In the fully turbulent region (Rey > 200) the RSM equations are used. In the viscosity-affected region (Rey < 200), the one-equation model of Wolfshtein [36] is adopted. Continuity between the two regions is provided by the blending equation of Jongen.

Combusting modeling

Fuel (CH_4/H_2) and air are premixed before entering the radiation section. The mixture composition used was 94.3% CH_4 and 5.7% H_2 by weight. The reaction scheme used, by Westbrook and Dryer (1981), was considered for the combustion of CH_4/H_2/ air mixture:

$$CH_4 + 1.5O_2 \rightarrow CO + 2H_2O$$

$$rCH_4 = 1.5x10^{13}e^{-125604/RT}C_{CH4}^{\ -0.3}C_{O2}^{\ 1.3} \quad (11)$$

$$CO + 0.5O_2 \rightarrow CO_2$$

$$rCO = 3.98x10^{20}e^{-167472/RT}C_{CO}C_{H2O}^{\ 0.5}C_{O2}^{\ 0.25} \quad (12)$$

$$H_2 + 0.5O_2 \rightarrow H_2O$$

$$rH_2 = 10x10^{13}C_{O2}^{\ 0.5}C_{H2} \quad (13)$$

IV. COMPUTATIONAL DYNAMIC FLOW MODEL

A computer model was developed using ANSYS Fluent package and CHEMKIN module.

Grid generation: the cross sectional grid was then extruded along the test tube centerline, i.e. a line for the straight tube and a helix for the swirl flow tubes.

Furnace operation and geometry condition: the pyrolysis furnaces function as both a reactor and high-pressure steam generator. It consists of six furnaces for cracking fresh and recycles feed. The pyrolysis reactions proceeds in tubular coils made of Cr/Ni alloys. Eight reactor coils, are hung vertically in a firebox, which is heated by means of 160 burners in total. They are arranged on the walls of the firebox for indirect firing. See Figure 2.

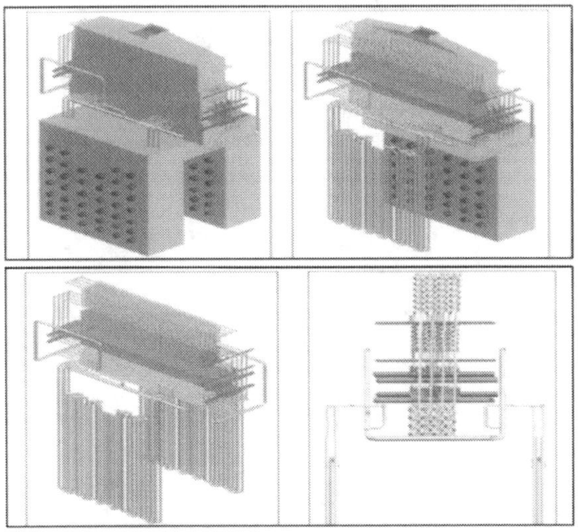

Figure 2. Isometric views of steam cracking furnaces

The main dimensions and operating conditions of the industrial cracking furnaces and adopted tubes are summarized bellow:

Table 1. Design and Operating Conditions for existing cracking furnace (industrial).

Inlet conditions	Ethane	Propane
Ethane/Propane Feed flowrate	40 T/h	52 T/h
Steam ratio (SR)	0,35	0,35
Temperature	695 °C	
Inlet Pressure	3.09 bar	

Table 2. Furnace dimension and operating conditions.

Furnace segment	
Height (m), (z -direction)	6.8
Length (m), (y -direction)	1.2
Width (m), (x -direction)	3
Number of burners	2
Burners diameter (m)	0.023
Material properties	
Emissivity of the furnace wall	0.85
Emissivity of tube skin	0.6
Thermal conductivity of refractory (W/mK)	0.394
Thermal conductivity of insulation (W/mK)	0.19
Thermal conductivity of tube skin (W/mK)	$-8.432+3.04\times10^{-2}T$
Reactor	
Tubes material	G-X40NiCrNb 35-25/ G-X40NiCrNb 35-25+ Ti+ Zr
Number of reactor tubes	8
Length (m)	78
Internal diameter (m)	0.1

Table 3. Geometry definition of adopted tubes

	Straight	SFT-M	SFT-H
Inside diameter(m)	0.1	0.1	0.1
Outside diameter(m)	0.18	0.18	0.18
Relative amplitude (-)		10.22	7.78
Helix angle (º)		7.63	13.5

III. RESULTS AND DISCUSION

1. Optimum operating conditions: they are vital to maintain efficiency, ethylene yield and operating cost:

Cracking severity: the product yields and selectivity are deeply affected by the growth of the coke layer and the amount of heat transferred to the reacting process gas and pressure level.

Furnace feedstocks: depends on the structure and molecular weight of the hydrocarbon.

Cracking temperature: the steam cracking reactions are highly endothermic; therefore high temperature increases the rate of formation of olefins and aromatics. However, high temperature also increases the rate of coke formation.

Partial pressure: to keep the reaction favorable towards ethylene, partial pressure needs to be reduced, thus, steam is introduced into the feed stream.

Residence time: affects the formation of aromatics and higher molecular weight hydrocarbons as a result of secondary reactions of formed olefins.

2. Operational failures:

Failures mechanism for radiant tubes: carburization and creep ductility exhaustion are presented in industrial cracking furnaces. Internal carbide formation results in bulging and ovalization of the tubes. It affects tube characteristics by impacting on stress rupture properties (circumferential cracking failure), ductility and thermal fatigue.

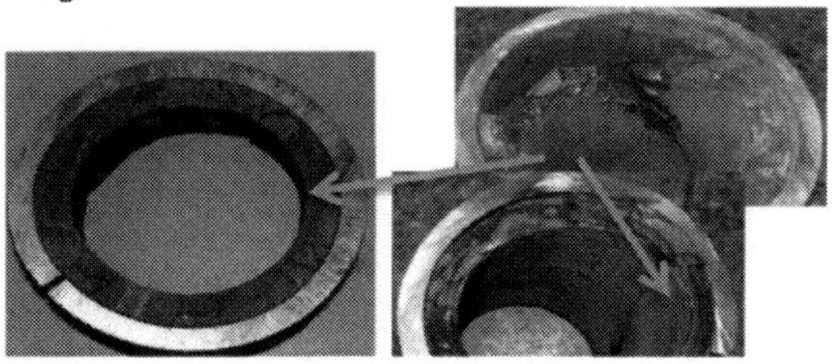

Image 1. Coke layer inside tube of the existing furnace.

3. Model validation

According to recommended tube heat flux and tube wall temperature profile presented by Sundaram and Froment [2] and cracked gas analysis of an industrially operated furnace (industrial data), model validation was done. Coil outlet temperature (COT), free radical concentration and ethane/propane conversion profiles along the reactor tube were obtained. Profiles were compared with industrial data and they showed agreement with them. These profiles are presented in Figures 3 and 4.

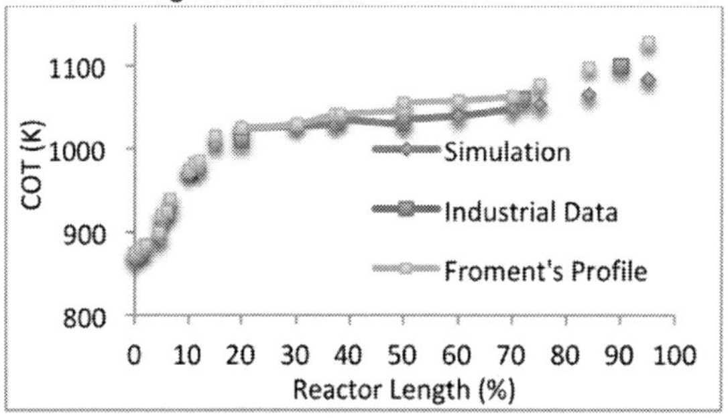

Figure 3. COT profiles. Simulation (CFD model), industrial data and Froment et al. Feedstock: Propane.

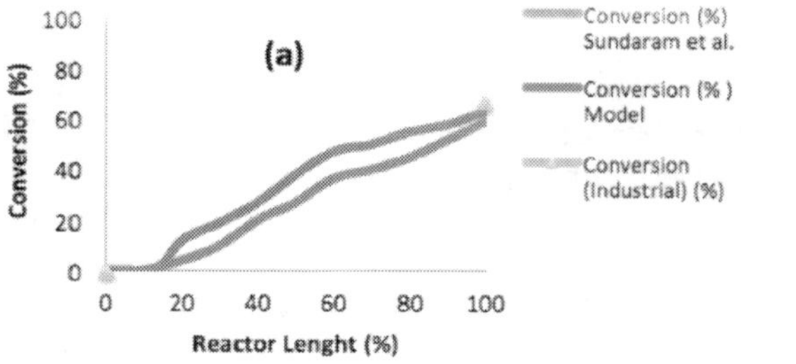

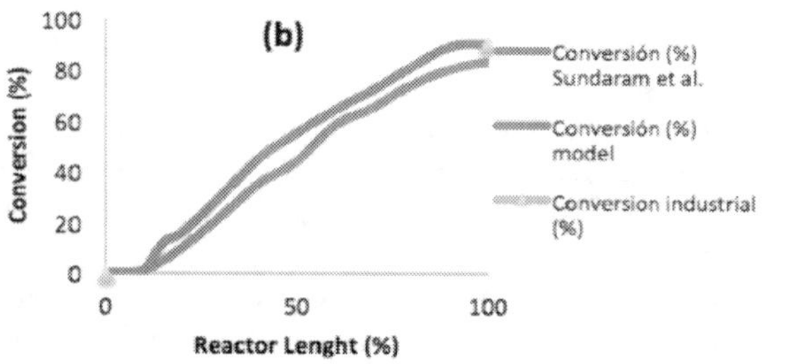

Figure. 4. Furnace conversion. Simulation (CFD), industrial data and Froment et al. (a) Ethane, (b) Propane.

4. SFT Technology Performance

The CFD model was adopted to capture the main flow properties of SFT technology. There were used three geometries: straight tubes (original design), mild swirl flow tubes (SFT-M) and high swirl flow tubes (SFT-H), considering ethane and propane as feedstock. To select best alternative, propane case was considered. This is an extreme condition that increases the pressure drop and reduces furnace selectivity and furnace operation length because of higher coke deposition.

The coke layer inhibits heat transfer from the tube to the process gas and hence is responsible for the rising tube wall temperatures. Operating parameters of the furnace such as pressure drop, conversion, velocity magnitude, heat transfer, and coke formation rate were evaluated:

Pressure drop: Fanning friction factors were evaluated considering Reynolds numbers from 30000 to 130000.

Based on these results, the Fanning factors are higher in propane feedstock. CFD model confirms that pressure drop is higher in SFT technology compared to straight tubes, similar to the results obtained from Schietekat, Albano et al and De Saegher experiments [1]. Pressure drop tends to decrease once fully-developed turbulent flow is established. Fanning friction factors in SFT-H tubes were higher than of the rest of geometries (0.013- 0.009).

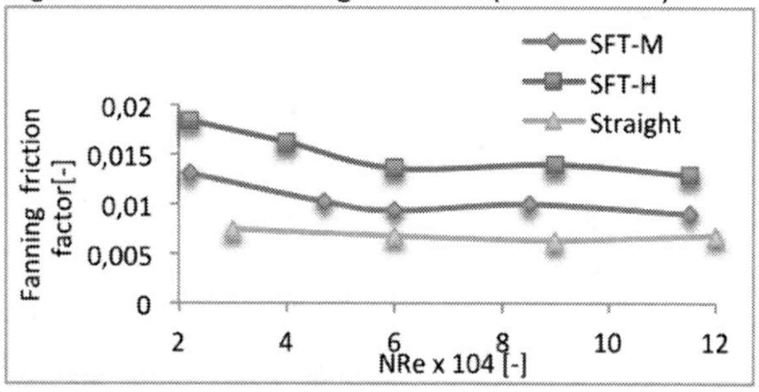

Figure. 5. Fanning friction factor. Simulation (CFD), industrial data and Froment et al. Feedstock: Propane.

Temperature distribution

Reactor temperature heavily influences the product distribution from the cracking process compared to steam to hydrocarbon ratio. Fig. 6 shows the radial temperature profile at the outlet of the tubes.

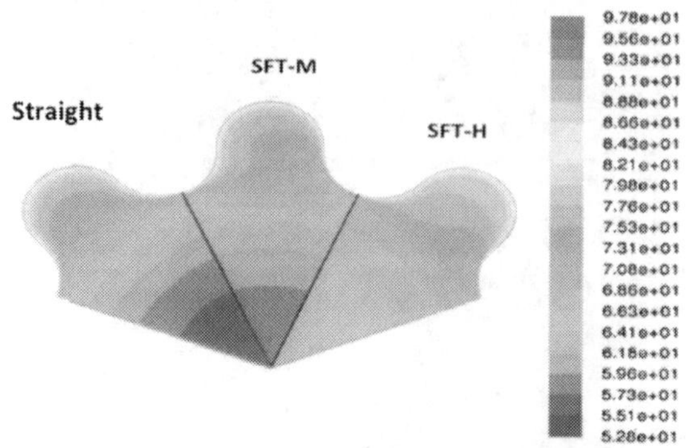

Figure 6. Temperature [°C] profile at the outlet of tubes Straight, SFT-M and SFT-H tubes. Feedstock: Propane.

SFT technology improves temperature uniformity in the coils of the reactor. The SFT-H tube shows best results, however as before mentioned at the expense of a higher-pressure drop. The more uniform temperature can be attributed to the stronger in-plane mixing by the swirl flow induced by the SFT and the higher heat transfer rate. As a greater temperature uniformity goes coupled with greater species concentration uniformity, a more uniform profile will provide lower coking rates and more controlled cracking (i.e. less under and overcracking). On the other hand, the area of high temperature close to the wall can induce a local high coke formation.

Selectivity: According to the endothermic process and increasing partial pressure in the cracking reactions, based on Le Chatelier's principle, by increasing the temperature and lowering the pressure, the conversion will increase in thermal cracking (Raseev 2003). See Fig. 5.

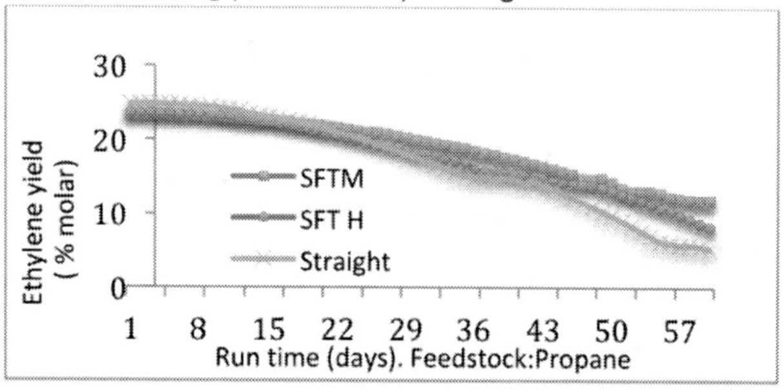

Figure 5. Evolution of ethylene yield in function of the run time

The reason for this drop in ethylene selectivity is made clear by the increasing of the coil inlet pressure in the cracking furnace. As the inlet pressure becomes higher, the hydrocarbon partial pressure will rise as well, leading to increased occurrence of the recombination reactions and as such lower ethylene yields. It can be seen that the ethylene yield for the straight tubes becomes smaller than SFT tubes after a certain run time. This can be explained upon consideration of the maximal thickness of the coke layer as a function of run time.

Coke formation

The increased heat transfer and pressure drop result from a higher wall shear stress. It is clear that the coke layer leads to reduce ethylene selectivity. While this is the case for all the coils, this drop appears to be more severe for the straight tubes. The evolution of coke layer in the tubes is shown in Fig. 6.

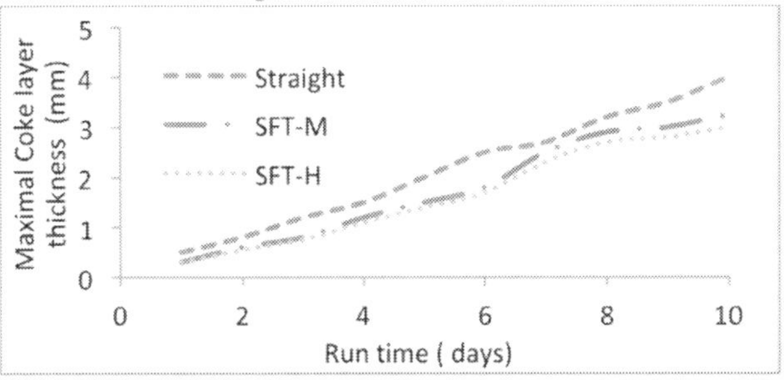

Figure 6. Evolution of the coke layer thickness as function of the run time. Feedstock: Propane

Comparing three different geometries, straight tubes have a more rapid coking rate and carburization, which shorten the tube-life.

Economic evaluation

It was carried out using Aspen Process Economic Analyzer, considering capital cost (CAPEX) and operating cost (OPEX) estimates. The industrial cracking furnaces used in this research present \$6.2 million per year losses in ethylene revenues and up to \$70 million per year in OPEX costs. SFT technology was evaluated to enhance cracking furnaces operation. Best criteria for selection were olefins production, coke formation rate and heat transfer. Incorporating SFT-M ("the best alternative") guarantees operational continuity, increases ethylene production of up to 13.5% (\$90MMA), reduces OPEX (\$10MMA) and raises furnace run length of up to 80 days without affecting mechanical integrity of coils. The SFT-M CAPEX contemplates \$577,823.00, which would be retrieved in olefins production in a period no longer than 6 months.

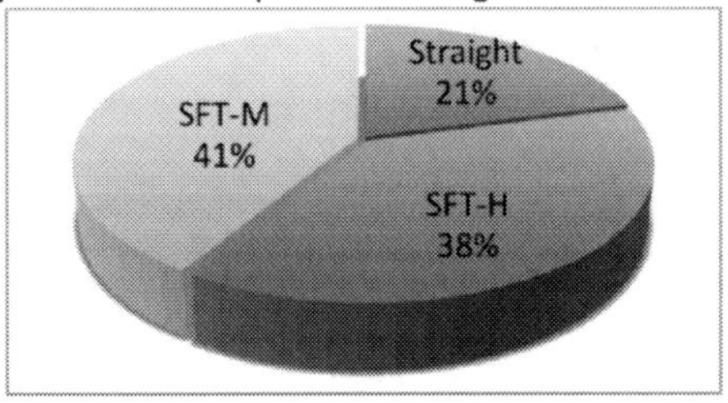

Figure 7. Benefits of using Straight/SFT-H/SFT-M tubes in cracking furnaces.

IV. Conclusion

Cracking furnaces of an existing olefins plant were used to study SFT technology due to its low production and operational failures because of coke formation leading damages to the material of the coils. Three geometries were used to evaluate feasible of this technology: straight tubes (original design), mild swirl flow tubes (SFT-M) and a high swirl flow tubes (SFT-H). The simulation results confirm the great potential for the application of the SFT technology.

Incorporating SFT-M ("the best alternative") would reduce \$0.6 per ton of ethylene in utilities costs and 17.5% in operational and maintenance costs compared to "straight" tubes. It increases furnace run length of 80 days and ethylene production of up to 13.5%. The SFT-M CAPEX estimates contemplates \$577,823.00, which would be retrieved in olefins production in a period no longer than 6 months period.

REFERENCES

[1] Schietekat, C., Van Goethem, M.W.M. Van Geem, K. y Marin, G.: "Swirl flow tube reactor technology: An experimental and computational fluid dynamics study". Chemical Engineering Journal In Press, Corrected Proof (2013). http://dx.doi.org/10.1016/j.cej.2013.08.086

[2] Sundaram K. M. y Froment G. F, "Modeling of thermal cracking kinetics III: Radical mechanisms for the pyrolysis of simple paraffins, olefins and their mixtures", Ing. Eng. Chem. Fundam, Volumen 17, Issue 3, 174-182 (1978).

[3] Sundaram, K.M., Van Damme, P.S. y Froment, G.F. Coke Deposition in the Thermal Cracking of Ethane. AIChE Journal, Volumen 27, No 6. (1981).

Nomenclature:

Roman

A= tube cross-sectional area, m^2

C= concentration, mol/m^3

CFD =computational fluid dynamics

COT= coil outlet temperature, K

C_p= heat capacity, J/mol K

d_t= tube diameter, m

F_T=molar flow rate, mole/s

f_f= Fanning friction factor

LMTD =log mean temperature difference

M_m= average molecular weight, kg/mol

Nu= Nusselt number

P= Pressure (Pa)

r= reaction rate in pyrolisis process, mol/m s

R_b= radius of the tube bend

Q= heat flux, W

Re= Reynolds number

SFT= Swirl Flow Tube

T= Temperature, (K)

X= axial length, m

x_1 = ethane mole fraction in the feed

x_2 = propane mole fraction in the feed

Greek

α_{ij} =stochiometric coefficients of component j in the ith reaction

Λ = Angle of bend (º)

ΔH =heat of reaction/mole

λ = thermal Conductivity (W/m/K)

ξ =Parameter of tube bend

ρ= Density (kg/m^3)

606177 The Challenge for Leaders Today

Abstract:

When women are asked the key characteristics of leadership, they tend to cite confidence and determination, but less than half of them feel that they have these characteristics. The challenge for leaders, then, is to instill these characteristics in others through mentoring relationships, especially to young women in academia. Training the next generation of leaders takes time and personal engagement to teach young colleagues how to plan their career, negotiate and find strong role models. Beyond gender considerations, leaders will find that issues surrounding communication and trust will play critical roles in their effectiveness at promoting an inclusive environment.

Copyright © American Institute of Chemical Engineers. All rights reserved.

602018 Building and Maintaining a Positive Culture in an Academic Department

Abstract:

Developing a departmental culture may sound like the epitome of a fuzzy, non-engineering soft skill, but the long-term differences between organizations with positive and negative cultures can be astonishing. I will reflect on tactics for building a culture that is positive for faculty, staff and students in an academic engineering department. In this environment, it is important to be sensitive to the time pressures faced by people at every level in the organization and the competitive nature of academic research and success.

Copyright © American Institute of Chemical Engineers. All rights reserved.

612621 Introductions of Panelists

Abstract:

Dr. Meredith Sellers is a Senior Managing Engineer in the Materials and Corrosion Engineering Practice at Exponent, an engineering and scientific consulting firm. As a Licensed Professional Chemical Engineer, she assists clients evaluate and investigate failures associated with material properties, processing, and performance. Dr. Sellers has significant experience assisting clients in the oil and natural gas and aerospace industries engaged in domestic and global legal disputes. She holds a B.S. and Ph.D. in Chemical Engineering from Cornell University and the University of Illinois at Urbana-Champaign, respectively. Dr. Sellers is a long-standing member of the Society of Women Engineers (SWE) and the Association for Women in Science (AWIS). She is involved in corporate initiatives within Exponent to both attract and retain women and underrepresented minorities in the workplace.

Dr. Felicia Etzkorn is Professor of Chemistry at Virginia Tech. She has a BS from Missouri State University, a PhD from the University of California, Berkeley, and was an NIH post-doc Harvard Medical School. Her research concerns the stabilization of proteins with small molecules to investigate cell division and collagen structure. She has been the sole PI on 3 major NIH grants, one NSF grant, and multiple smaller grants. She participates in two major NIH biomedical training grants for underrepresented students. She created a Green Chemistry course for senior undergraduates at VT in 2007. Her textbook, *Green Chemistry: Principles and Case Studies* was published in January 2020.

Motunrayo Kemiki is a Renewable Energy and Project Management Professional with more than 10 years experience in the federal government. She currently works at the Office of Renewable Energy Programs at the Bureau of Ocean Energy Management (BOEM) where she is responsible for managing reviews of technical plans submitted for offshore renewable energy projects in the Outer Continental Shelf. Prior to joining BOEM in 2014, she worked as a Project Manager at the Department of Energy Loan Program Office (where she managed technical due diligence of multi-million dollars renewable energy projects -biofuel, energy efficiency & transmission-under Title XVII of the Energy Policy Act) and as a project engineer at the Naval Surface Warfare Center (where she coordinated the manufacturing of rocket motor propellant for the MK6).

She has a Bachelor of Science in Chemical Engineering from the Pennsylvania State University and a Masters in Business Administration from the University of Maryland, College Park. She is also a licensed Project Management Professional.

Named one of the 100 Women Leaders in STEM by STEMconnector, **Tricia Berry** leads efforts to recruit and graduate women in the Cockrell School of Engineering as Director of the Women in Engineering Program at The University of Texas at Austin. She concurrently serves as Collaborative Lead for the Texas Girls Collaborative Project (TxGCP), leading the dissemination of STEM best practices and informal curriculum across Texas in coordination with the National Girls Collaborative Project. Through both roles, she connects and supports organizations and individuals working to advance gender equity in science, technology, engineering and math (STEM) fields across Texas and beyond.

Tricia is the co-author of You Can't Eat Your Degree: Combine Your Passions and Philosophies to Create the Story of Your Future and Exceeds Expectations: Take Control of Your Performance Review. Previously, Berry worked as a Process Engineer and a Product Development Engineer at The

Dow Chemical Company in Freeport, Texas.

Dr. Jessie DeAro is an expert in broadening participation in STEM education research and diversity programs and has nineteen years of experience designing, implementing, and leading federal grant programs at the National Science Foundation (NSF) and the U.S. Department of Education. She has the managed several different diversity related grant programs at the NSF and is currently the NSF ADVANCE program lead and a program director with the EHR Core Research and the Science and Technology Centers programs. She has served in leadership as acting Deputy Division Director at NSF and in policy roles as a senior policy advisor at the White House Office of Science and Technology Policy in 2010.

Ekin O. Gozen is currently the Strategic Portfolio Advisor at Shell located in SF Bay Area. In her current role, as part of Technology Portfolio team, Ekin supports portfolio management and decision making for the Vice President of Refining, Catalyst and Analytical technologies. She delivers cross-portfolio analysis and insights on Shell Technology strategy implementation, productivity of technology portfolio, alignment of portfolio to business technology needs for CTO, the Technology Leadership Team and other key stakeholders. Prior to her current role, Ekin held various roles at Shell Martinez Refinery including the Team Lead for the Quality Assurance group and Process Engineer at Projects team. Ekin joined Shell in 2013 as a Research Engineer in Process Evaluations where she provided technical and economical guidance to various Shell businesses such as Alternative Future Energy, Chemicals, CO_2 management and Upstream. Ekin holds a PhD in Energy Resources Engineering from Stanford University and MSc and BSc in Chemical Engineering from Middle East Technical University in Turkey. In her spare time, Ekin enjoys singing with her band, running and cycling.

Copyright © American Institute of Chemical Engineers. All rights reserved.

AUTHOR INDEX